AGROFORESTRY DEVELOPMENT ON THE CANADIAN PRAIRIES

AGRICULTURE ISSUES AND POLICIES

Manure Use for Fertilizer and Energy
Connor D. Macias (Editor)
2010. 978-1-60876-847-9

Price Dynamics Behind Consumer Food Purchases
Morgan D. Fitzpatrick (Editor)
2010. 978-1-60876-892-9

Governance of Agrarian Sustainability
Hrabrin Bachev
2010. 978-1-60876-888-2

Transformation of U.S. Animal Agriculture
Justin M. Daigle (Editor)
2010. 978-1-60876-938-4

The Peanut Plant and Light: Spermidines from Peanut Flowers and Studies of their Photoisomerization
Victor S. Sobolev, James B. Gloer and Arlene A. Sy
2010. 978-1-61668-028-2

Agriculture and Environmental Security in Southern Ontario's Watersheds
Glen Filson
2010. 978-1-61668-156-2

Million Dollar Farms in the New Century
Samuel D. Bosworth (Editor)
2010. 978-1-61668-541-6

Agriculture and Environmental Security in Southren Ontario's Watersheds
Glen Filson, Bamidele Adekunle and Katia Marzall
2010. 978-1-61668-372-6

Agroforestry Development on the Canadian Prairies
Suren N. Kulshreshtha, Ken Van Rees, Hayley Hesseln, Mark Johnston, John Kort
2010. 978-1-61668-588-1

Alignment-Free Models in Plant Genomics: Theoretical, Experimental, and Legal issues
Humberto González-Díaz, Guillermin Agüero-Chapin, Cristian Robert Munteanu, Francisco Prado-Prado, Kuo-Chen Chou, Aliuska Duardo-Sanchez, Grace Patlewicz and Antonio López-Diaz
2010. 978-61668-333-7

The Peanut Plant and Light: Spermidines from Peanut Flowers
Victor S. Sobolev,James B. Gloer and Arlene A. Sy
2010. 978-1-61668-371-9

Manure: Management, Uses and Environmental Impacts
Carmen S. Dellaguardia (Editor)
2010. 978-1-61668-424-2

Agricultural Economics: New
Research
Tomas H. Lee (Editor)
2010. 978-1-61668-445-7

Alignment-Free Models in
Plant Genomics: Theoretical,
Experimental, and Legal issues
*Humberto González-Díaz,
Guillermin Agüero-Chapin, Cristian
Robert Munteanu, Francisco Prado-
Prado, Kuo-Chen Chou, Aliuska
Duardo-Sanchez, Grace Patlewicz
and Antonio López-Diaz*
2010. 978-1-61668-603-1

Manure: Management, Uses
and Environmental Impacts
*Carmen S. Dellaguardia
(Editor)*
2010. 978-1-61668-647-5

Agriculture Issues and Policies

AGROFORESTRY DEVELOPMENT ON THE CANADIAN PRAIRIES

SUREN N. KULSHRESHTHA
KEN VAN REES
HAYLEY HESSELN
MARK JOHNSTON
AND
JOHN KORT

Nova Science Publishers, Inc.
New York

LIBRARY OF CONGRESS CATALOGING-IN-PUBLICATION DATA

Available upon Request
ISBN: 978-1-61668-266-8

Published by Nova Science Publishers, Inc. ✢ *New York*

CONTENTS

PREFACE

Agroforestry is a system of land use in which both agriculture and forestry production are intermingled. All over the world, farmers have had a long tradition of planting trees on their fields and pastures. The major objective of this book is to review the history of agroforestry development in the region and this progress is found to be slower than in other regions of North America in spite of various benefits that could be generated from agroforestry practices. This book reviews the state of agroforestry in the Prairie Provinces of Canada and describes various types of agroforestry that have developed historically.

1. INTRODUCTION

Agroforestry is a system of land use in which both agriculture and forestry production are intermingled. All over the world, farmers have had a long tradition of planting trees on their fields and pastures. Using the internationally accepted definition of agroforestry, the Saskatchewan Advisory Council on Agroforestry (SACAF) defined it as "a land-use system that combines trees and/or shrubs with agricultural plants and/or animals on the same land area." (Government of Saskatchewan, 1997). Under temperate systems (such as in the Canadian Prairies), agroforestry can include various types of practices. Major ones include: alley cropping, forest farming, shelterbelts (or timberbelts), riparian buffer strips, and silvopastoral systems (Garrett et al. 2000).

This study reviews the state of agroforestry in the Prairie Provinces[1] of Canada and describes various types of agroforestry that have developed historically. The major objective of the study is to review the history of agroforestry development in the region. This progress is found to be slower than in other regions of North America in spite of various benefits that could be generated from agroforestry practices, which are reviewed here. Reasons for slower progress are cited and policies that may lead to producers adopting such practices on prairie farms are listed.

---

[1] The Prairie Provinces include three provinces located in western half of Canada: Alberta, Saskatchewan, and Manitoba.

2. HISTORY OF AGROFORESTRY IN THE PRAIRIE PROVINCES

The Prairie Provinces have had a long, storied history with tree planting starting back in the 1870s, some 60 years before the tree planting expansion of shelterbelts to control severe soil drifting during the Great Depression. Although the planting of trees for afforestation or shelterbelts was a common occurrence back then, it was not known as agroforestry[2]. In 1870, the North-West was transferred to the Dominion of Canada. In order to encourage settlement in the west, the federal government felt that a tree planting program would enhance the landscape, produce more rainfall and aid in successful agriculture. The west was considered very arid but there had been reports from various scientific expeditions that portions of the prairies were very fertile and that a large scale tree planting endeavour (up to 1/3 of the region) would create suitable conditions for prairie farming (Watters, 2002). In addition, eastern Canada was undergoing large scale deforestation and there was concern that there might be a shortage of lumber – hence planting of the prairie region would provide further sources of timber as settlement developed westward. There were also theories that large-scale plantings of trees would increase rainfall which would enable successful farming. Thus from a period of 1870 to 1886, afforestation (tree planting on a large-scale) was being promoted on the prairies. To assist the settlers with their farm practices, including tree planting, the federal government passed the Experimental Farms Act in 1886, which created Experimental Farms that would do research on tree planting and

---

[2] In this chapter it is assumed that these producers were practicing a form of modern day agroforestry, a term that was not used until 1977.

growth, extension and provide trees to the settlers. Experimental farms were established in Brandon, Manitoba and Indian Head in the North-West Territories, as well as in Ottawa. In 1888, the farm in Ottawa sent 20,000 trees to Indian Head and in 1889 a massive trial of 27,000 trees and shrubs from 718 varieties were sent to Brandon and Indian Head farms to determine their hardiness in the prairie climate (Howe, 1986). By 1901, the Forestry Branch of the Department of the Interior began propagating plant material at Brandon and Indian Head and in 1902 began a program of providing trees to farmers free of charge (Howe, 1986). In 1902 a permanent nursery was built in Indian Head and in 1963 the nursery was transferred to the Prairie Farm Rehabilitation Administration (PFRA) under Agriculture Canada[3]. However, during the period 1887 to 1914, it became evident that there was not enough planting stock for the engineering of large landscape plantings, since the settlers were more interested in planting trees around their farms than being involved in establishing plantations. Thus from 1887 to 1914, the emphasis by various government agencies and farm and forestry advocate groups shifted to the planting of shelterbelts to protect settlers' homes, gardens, crops and livestock and as a source of fuel and fence posts.

By 1883, when the Canadian Pacific Railway (CPR) extended across the prairies, some settlers had started importing tree seedlings from Eastern Canada. Unfortunately, they were generally unsuccessful in their plantings, mainly because they were using non-hardy species (Ross, 1923). For many farmers, it became general wisdom that it was impossible to grow trees on the prairies. However, some persevering farmers, using native species such as Manitoba maple, green ash and American elm, were able to demonstrate the successful development of shelterbelts. When it became clear that trees for shelter and fuel wood could be successfully established, the Government of Canada, in 1901, developed what Ross (1923) called "the co-operative system". Understanding that the settlers wanted and needed the trees, but that they did not have the money to buy them, the Canadian Government began, in 1901, to distribute cuttings, seedlings and seed to landowners who promised to follow established regulations – to prepare the soil the year before planting, to avoid planting trees within 27 m of any permanent buildings and to follow species and spacing recommendations for planting design. The material was provided for forest plantations and shelterbelts (Department of the Interior, 1910). The arrangement was co-operative in that the government provided the

---

[3] The name of PFRA was changed in 2009 to Agri-Environment Service Branch – AESB. In subsequent discussion this will be referred as AAFC-AESB.

material but the farmers provided all of the labour and other input costs like land, farm equipment and fencing.

The need for fuel wood plantations was not general throughout the prairies since there was abundant natural wood in the parkland area and some prairie river valleys were well-treed. Even prior to 1892, ox trains were freighting coal from the Estevan area, where it had been discovered in 1857 by Captain John Palliser's expedition (MacKenzie, 2003). In 1892, a railroad track connected Estevan to the CPR mainline resulting in reduced need for fuel wood, even though some landowners were still interested in tree plantations for fence rails and rough lumber.

The real need for planting trees was for shelterbelts, although many farmers had become resigned to a wind-swept existence on the plains. At the start of the Canadian Government's co-operative program, the uptake in 1901 was only 58,000 seedlings by 47 farmers, but this quickly increased to 1.8 million by 1904 and to 7.7 million by 1917 (Ross, 1923). Most of the planting at first was done around farmyards and farm houses. Farmyard shelterbelts continue to be planted to the present day, with a result that there are few farmyards throughout Manitoba, Saskatchewan and Alberta that do not have shelterbelts surrounding them or, at least, protecting them from the predominant or problem wind directions. By 1914, planted trees had outnumbered the settlers on the prairie landscape 30 to 1 which altered the view of the once tree-less prairies. A list of trees and shrubs that were available for planting according to suitable soil type are listed in Table 1. Norman Ross, who was the first superintendent at the Tree Nursery in Indian Head (Howe, 1986), summarized all the research and information related to tree-planting in a publication called 'Tree-planting on the prairies of Manitoba, Saskatchewan and Alberta' (Ross, 1910). Other guides for tree planting on the prairies were published by Ross (1939) and Ellis et al. (1945).

Some pioneering farmers scattered throughout the prairies began more systematic field shelterbelt plantings to protect their fields and crops. Ross (1923) reported that, while the most of 40,000 shelterbelts established by 1923, were planted around buildings and gardens for protection, there was an increase in farmers who planted shelterbelts to prevent soil-drifting. Then the "Dirty Thirties" arrived, when the economic depression coincided with years of drought, grasshopper infestations and soil erosion. With government assistance, those farmers who farmed the more erodible prairie soils planted many miles of field shelterbelts, which provided them short-term income, as well as a way of providing long-term soil protection and trapping much-needed moisture in the form of snow. Since that time, field shelterbelts continue to be planted based on farmers' needs and circumstances. Prolonged drought in the 1980's resulted in the greatest annual number of field shelterbelts being established in 1990. Although years with normal precipitation would be recommended for best establishment and growth, the reality is that tree planting increases when it suddenly becomes clear that the lack of trees is a problem.

Table 1. List of trees suitable for planting on different soil types (Source: Ross, 1910)

Heavy Clay	Moist, Sandy Loam	Dry, Sandy Loam	Sand or Gravel	Low, wet land
Man maple	Man. maple	Man. maple	Russian poplar	Ash
Soft maple*	Soft maple	Russian poplar	White spruce	Elm
Scrub oak*	Green ash	some willows	Scotch pine	Cottonwood
Basswood*	Basswood	Scotch pine	Jack pine	Black poplar
Green ash	Elm	Jack pine		Larch
Elm	Cottonwood	White spruce		Black spruce
Cottonwood	Willow			Willow
Willow	Birch			
Larch	Larch			
Scotch pine	Scotch pine			
	Jack pine			
	White spruce			

* These trees are suitable for only portions of southeastern Manitoba.

3. THE NATURE OF AGROFORESTRY ON THE PRAIRIES

Historically, most agroforestry in the Prairie Provinces has taken the form of shelterbelts (Mak et al. 1999), although there are now new production systems and tree species and varieties that are making a wide variety of agroforestry practices economically promising. Shelterbelts flourish across the Prairie Provinces, and have been used since the 1930s primarily as a means of reducing erosion, yet not as a crop to be managed and harvested for profit. Today, farmers are exploring additional methods of introducing or incorporating woody crops into existing production systems including shelterbelts, but expanding to intercropping, silvopasture, fruit and oil production, and ornamentals. In this section different types of agroforestry are described.

3.1. SHELTERBELTS

Shelterbelts are linear arrangements of trees or shrubs. They perform a variety of functions that benefit ranching and farming operations and are used to: reduce wind speed; reduce soil erosion; increase local soil moisture by trapping snow; protect crops; protect biodiversity and create habitat; create a sound buffer; create a sense of privacy on the prairies; provide shelter for livestock, farm infrastructure and farm families; and they are aesthetically pleasing and may increase land value.

The use of shelterbelts on the prairies has evolved with the circumstances and needs of prairie farmers. Currently, shelterbelts continue to be planted on the Canadian prairies for the same reasons as they were fifty years ago, but increased experience with shelterbelts and greater knowledge of their effects, designs and limitations mean that their use is perhaps more targeted than in the past. There has been an increased recognition of additional benefits of shelterbelts or other agroforestry plantings on the prairies, which has involved the participation of a variety of environmental groups and initiatives in tree-planting projects. To the extent that landowners are able to economically benefit from the greenhouse gas reduction from their shelterbelt-planting activities, this is likely to increase shelterbelt use and shelterbelt-planting activities.

There are also reasons for a decrease in shelterbelt use. Although the onset of a drought cycle could trigger another increase in field shelterbelt planting, advances in direct-seeding technology have led many landowners to conclude that they can manage soil erosion without the use of shelterbelts. Meanwhile, larger field equipment and the increased trend to farming rented land may also be reasons for recent decreases in field shelterbelt planting.

3.2. Tree Farming or Plantations

Plantations or tree farming has great potential in the Northern Prairies as there are large areas of land that could be planted to trees relative to other agricultural land in Canada. This concept was promoted in 1904 as farm forestry where settlers were encouraged to plant 10 to 15 ha in windbreaks to supply firewood and fencing material on their farms (Mitchell, 1904). These planted areas were to be divided into equal units that could be cut every year for a certain rotation until the first area could be harvested again because the trees were self-propagating. Thus these early beginnings of 'sustainable' forest farming are still quite applicable today.

Plantations today are predominately focussed on fast growing species such as hybrid poplar. The major operational plantings of hybrid poplar are occurring in north-eastern Alberta as part of Alberta Pacific Forest Industries (Al-Pac) plan to supply a portion of their wood fibre for their pulp[4]. A number of hybrid poplar trials have been established in Saskatchewan to investigate the role of tree spacing, fertilization effects and clonal trials (Booth, 2008).

---

[4] More detailed information on the Al-Pac plantings is reported in section 8.1.

Photo 1. Hybrid poplar plantations planted on an alfalfa field near Meadow Lake Saskatchewan. Plantations on the left are the spacing trials and in the foreground is the clonal X fertilizer trial. (Photo courtesy of Mistik Management Ltd.)

The oldest trial (planted in 1997) is a 12-year-old spacing trial of Walker poplar (*P. deltoides X P. Xpetrowskyana*) in the aspen parkland region which has yields of ~100 m^3 ha^{-1} for 2.4 X 2.4 m spacing. Mean annual increments range from 8 to 9.5 m^3 ha^{-1} yr^{-1} and tree growth models have predicted that these stands may yield up to 220-250 m^3 ha^{-1} of wood by age 20 (unpublished data). Hybrid poplar has been the focus as its wood fibre is suitable for pulp and oriented strand board as well as engineered wood products (e.g. laminated veneer lumber) and wood flooring. A poplar breeding program at AAFC-AESB Agroforestry Development Centre in Indian Head currently produces about 7 clones (Assiniboine, Hawktree[5], Hill, Katepwa, Northwest[6], Okanese and Walker) which have been released for the Prairie Provinces through its breeding program (AAFC-PFRA, 2007). An active breeding program is needed to improve hybrid poplar for drought and cold tolerance as well as disease resistance. Few diseases have been identified in these plantations although septoria canker has been identified in some plantations in the eastern

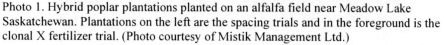

[5] Not yet available for distribution.

[6] This clone was developed in North Dakota.

portion of Saskatchewan[7]. Other plantations of conifer species, such as red pine, Scots pine and Siberian larch, have been established in recent years, although these species were tested in the early 1900s as well and some of these plantations still exist today at the AAFC-AESB Agroforestry Development Centre. White spruce is also a sought after tree species for lumber and could perform well on agricultural soils in the aspen parkland area and throughout the prairies. However, other high-valued species that are comparable to the walnuts or cherry trees of eastern North America such as white birch and green ash need to be examined for their potential as a marketable species on the prairies. The location of current plantations in Saskatchewan is presented in Figure 1 and details for each plantation have been documented by Letwinetz and Jurgens (2008). The total area planted to 2008 is approximately 500 ha and specific information about types of plantations is presented in Table 2.

Table 2. Summary of total planted agricultural area in Saskatchewan to December 2008

Conifer plantations total	59.9 ha
Conifer/hybrid poplar plantations total	91.4 ha
Hybrid poplar plantations only total	329.3 ha
Balsam poplar plantations total	2.3 ha
Willow (biomass) plantations total	5.4 ha

Source: Information compiled by ForestFirst (2009).

Plantation management is a key factor in the successful establishment of planted trees in prairie soils. Soil cultivation is required the year prior to planting to minimize vegetation competition and a pre-emergent herbicide is necessary at the time of planting to control weeds. Weed management is generally required for three to four years after which crown closure will shade out any weed competition. One of the major issues with herbicide application is that there are very few registered for use in tree plantations; many are registered for trees in shelterbelt systems but not for tree farming. The Pesticide Working Group of the Poplar Council of Canada has undertaken the process of registering new herbicides with the Pesticide Management Regulatory Agency (PMRA). It must be stressed, however, that without

---

[7] Although hybrid poplar is a fast growing species and suitable for planting in agricultural soils, emphasis must be placed on developing other species for different products and markets.

satisfactory vegetation control, plantation establishment and successful growth will not be realized and could result in the loss of the initial investment.

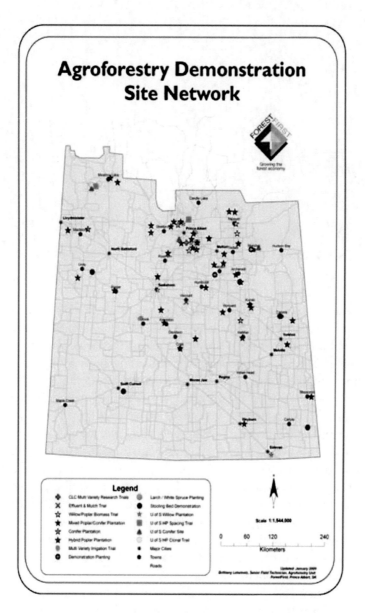

Figure 1. Location of Agroforestry Demonstration Sites in Saskatchewan.

Photo 2. A 12 year old plantation of Walker hybrid poplar grown on agricultural land.

The establishment of stooling beds are necessary for providing cuttings for planting and consist of closely planted rows of material that are coppiced each year and cut into desired lengths for planting. A number of stooling beds have been established across the prairies where planting stock for hybrid poplar can be obtained. Studies have shown that of the three different types of planting stock available to farmers, rooted material (root cuttings from nursery beds or rooted plugs from tree nursery greenhouses) have better survival compared to cuttings. Although cuttings are cheaper to purchase, the risk of not achieving successful establishment is greater for successful establishment especially in dry spring conditions. Studies of fertilization at time of planting have indicated that hybrid poplar trees and conifers do not respond to nitrogen applications during the first five years after planting. Whether the agricultural soils have high enough fertility from previous fertilization, crop rotations or inherent soil fertility is unknown at this time. Thus, fertilization of tree plantations at time of establishment is usually not needed; however, mid-rotation fertilization will likely be required for longer rotations.

With climate change being an important consideration and annual water deficits the norm for the prairie region, the possibility of irrigation for these crops needs to be examined. Currently a study is exploring irrigation trials in Outlook, SK, with various fast growing woody species. The application of wastewater effluent from cities, rural municipalities or hog barns onto plantations or shelterbelts has the potential to benefit growth as well as provide an environmentally friendly and economical approach for disposing of waste.

3.3. INTERCROPPING (ALLEY CROPPING)

Intercropping is a term that has been used interchangeably with alley-cropping to describe agroforestry systems in which crops are combined with closely-spaced rows of trees or shrubs with narrow alleys between them. Combining trees with adjacent agricultural or horticultural crops in the wide alleys can involve high value hardwoods or fast growing species such as hybrid poplar (University of Minnesota Extension, 1999). The same biophysical interactions occur as with traditional shelterbelts, but they are more pronounced and management and equipment needs are different than for large fields. In general, intercropping has not been adopted by farmers in western Canada but has been used in particular situations, such as when the crops in the alleys are especially valuable or when the trees or shrubs themselves also have significant value.

There are examples of successful prairie intercropping systems. You-pick strawberry farms, in which the strawberry crops are grown in the alleys, often use hedgerows to provide protection for the strawberry plants and the fruit at all stages of their growth. The shelter ensures even snow cover that protects strawberry plants in the winter, improves fruit yield and quality and also assures the comfort of clients who come to pick the fruit. In some cases, growers have invested in artificial windbreak nets or fences, while others use saskatoon berry bushes for the windbreaks, which provide another valuable crop while providing shelter for the strawberries. Some vegetable crops such as tomatoes, peppers and melons produced by market gardeners also improve in yield and quality from such shelter, although the increased temperatures may cause earlier flowering of species such as lettuce, radishes or spinach. Commercial nurseries for woody and perennial plants normally ensure their material is well-protected because the value of the plants is directly related to their visual appeal and quality.

The most recognized intercropping trial in Canada is the long-term tree-based intercropping trial that was initiated in 1987 at the University of Guelph, Guelph, Ontario. The trials incorporated 10 different tree species [*Acer saccharinum* (silver maple), *Corylus avellana* (hazelnut), *fraxinus americana* (white ash), *Juglans nigra* (black walnut), *Picea abies* (Norway spruce),

Populus sp. (poplar –hybrid), *Quercus rubra* (red oak), *Robinia pseudoacacia* (black locust), *Salix discolor* (willow) and *Thuja occidentalis* (white cedar)] that are spaced either 12.5 or 15 m apart and annually intercropped with maize (*Zea mays*), soybean (*Glycine max*), and winter wheat (*Triticum aestivum*) or barley (*Hordeum vulgare*) (Thevathasan and Gordon, 2004). Today, researchers at the University of Guelph have altered the intercropped species by fast growing willow as a biomass species in the inter rows on a three year coppice rotation.

Although intercropping systems at these narrow row spacings would be impractical for prairie farmers using large machinery, some intercropping systems are being studied on the prairies. The AAFC-AESB Agroforestry Development Centre has been investigating the use of willow as an intercropping system. The plantings in this study have been designed so that closely-spaced willow hedgerows could be regularly coppiced in rotation for their biomass, while the remaining hedgerows would continue to provide shelter for the intercrops. Other plantings, with poplars as the tree rows, have been established in Manitoba and Saskatchewan with the idea that the high-nutrient requirements of the poplars could be met by applications of swine effluent, while helping the trees to maximize their growth, so that they would be valuable for producing high-quality, clear poplar wood. Between tree rows, annual or perennial crops are grown.

3.4. SILVOPASTURE

Silvopasture involves the incorporation of livestock, pasture and trees into a farming system. This practice is used extensively around the world (Gordon and Newman, 1997) and has been discussed in a Saskatchewan context by White (2005) and Kort et al. (2008). Grazing livestock in aspen or other forested stands in the prairie parkland is a common practice and cattle benefit from the presence of the trees because of the shade and shelter that they provide. However, forage production under forest stands is much reduced because of shade and competition, while the trees in some of these forests are economically desirable for commercial use as pulp, chips or lumber. When commercial forests are not grazed or grazed forests are not commercially harvestable, they cannot be considered to be silvopasture because there is no aspect of co-management of the tree and agriculture components. In general, silvopasture practices have not been adopted in the prairies, perhaps because of the relatively low value of wood products at this time. However, under favourable economic circumstances, there clearly is potential for this practice because of the large livestock industry and the extensive land area under forest.

Photo 3. An example of a silvopasture trial where trees were planted into forage and protected by electrical fence.

3.5. OTHER – RIPARIAN BUFFERS

In addition to the types of agroforestry described above, other examples can also be noted. In particular, low shrubs near wetlands provide excellent nesting cover for waterfowl, while multi-row, multi-species upland shelterbelts provide habitat for plants, insects, birds and mammals. Riparian buffers that include trees and shrubs can protect aquatic habitat as well as prevent agricultural nutrients and sediments from degrading water quality, thereby involving watershed management organizations in tree-planting projects. Shelterbelts especially designed for protecting laneways, roads and highways from snow have been planted by individual landowners, municipalities and provincial highways departments. Current concerns about climate change and greenhouse gas emissions have resulted in a new look at shelterbelts as a way to sequester carbon or as a way to sustainably produce woody biomass as a renewable fuel.

Many examples of successful agroforestry can be identified in the Prairie Provinces. One such example of the Parkland Agroforestry Incorporated is shown in Box 1.

BOX 1

Parkland Agroforestry Incorporated – a Prairie Success Story

 Parkland Agroforestry is a group of private landowners situated in northeast Saskatchewan who work together to develop new opportunities in agroforestry. The group was founded in 2000 and works to collectively provide agroforestry education and resources for its members, research and development opportunities for new crops, and support in working with government to develop policy in support of an agroforestry industry. Parkland's success is based in their ability to identify niche markets and actively pursue development opportunities through research, and to assess business and trade relationships.

Group members are located close to the boreal forest where growing conditions support production of aspen, poplar, white spruce and jack pine. While these species are well suited to producing commodities such as lumber and pulpwood Parkland has diversified and engaged in active management and planning that includes developing hybrid poplar for stooling beds and greenhouse stock, willow and haskap (pictured above).

As an example, haskap, also known as blue honeysuckle, is highly desired for pastries, jams, juice and wine, to name a few, particularly in Japan where it is prized as a native plant. It is a hardy and long-lived plant, ideally suited to prairie growing conditions. Parkland Agroforestry has worked with researchers from the University of Saskatchewan (see Lefol, 2007; Bors, 2008) to assess market potential and competition in entering into this market.

Although agroforestry in the Prairie Provinces has primarily been limited to shelterbelts, there are new opportunities emerging that serve to help farmers diversify and to take advantage of niche markets. In so doing, such producers develop new economic opportunities to increase profits, but also provide public goods by sequestering carbon, producing non-market benefits derived from habitat production and carbon sequestration for example.

For more information see Parkland Agroforestry at: www.parklandagroforestry.com

4. BENEFITS OF AGROFORESTRY DEVELOPMENT

Agroforestry has the potential to benefit society, in addition to private landowners. The value to society of various types of agroforestry (particularly from shelterbelts and plantations) stems from various functions, many of which are ecological.

Benefits derived from agroforestry can be divided into two types -- private and public. Private benefits accrue to the farmer or rancher and will provide direct profits from harvesting or use of the crop, and indirect profits from enhanced moisture retention and therefore, greater returns to crops, enhanced livestock values in the case of silvopasture, and enhanced water quality in riparian areas. Producers, and members of society, may also gain non-financial rewards in the way of non-market benefits. These could include enhanced aesthetics or increased wildlife activity, for example. In the latter case, such benefits are considered public and also accrue to society in general. Other non-market or public benefits could include reductions in soil erosion, increases in wildlife habitat, enhanced aesthetics, land reclamation, waste management, etc. In this section, both private and social benefits of agroforestry are described, with a particular emphasis on shelterbelts.

4.1. PRIVATE BENEFITS

While shelterbelts have been widely planted across the prairies, other agroforestry practices and tree plantations have not been widely adopted, given the abundance and accessibility of trees on crown lands in the northern part of the provinces. But new opportunities are arising for farmers to incorporate woody crops to diversify financially and to produce non-market or social and public benefits (Mak et al. 1999). This is particularly pertinent when it comes to carbon sequestration as it is related to climate change. Although more opportunities are being presented for private benefits, historically such benefits were recognized by producers. For example, in 1910, landowners receiving trees through the shelterbelt program were invited to provide their comments on the values of their shelterbelts, (Department of the Interior, 1910). The report included comments such as the following:

> "... there is not the slightest doubt that a well-established plantation adds very many dollars to the sale value of the farm."
> "The presence of trees around the farmstead makes all the difference between a 'home' and a 'house', between 'living' and merely 'existing.'
> "It is practically impossible to have a really comfortable and attractive home without the shelter and adornment which is afforded by the liberal planting of trees."

Letters from 66 farmers in Alberta, Saskatchewan and Manitoba estimated that their shelterbelts, young as they were, had values of over $1000, when adjusted for inflation would approximate $23,000 today. Some letters contained comments such as:

> "We would not be without them now for anything." and "... money could not induce me to remove them."

Although farmers might express their sentiments differently now, the shelterbelt program continues to be well subscribed to, with the 600 millionth tree sent out under the program planted in 2008. Although the trees may not have market value in terms of producing marketable wood, the inputs that landowners invest into the shelterbelts is substantial, including the opportunity cost of the land used for the shelterbelts, the labour, chemical, machinery and fuel inputs.

There are also economic incentives for planting trees, although as a means of economic diversification, revenues will vary by crop and production system. Factors that affect profitability include location and land quality; the availability of processing infrastructure and the distance to mills in the case of woody crops used as traditional forestry output (i.e. dimension lumber, veneer, oriented strand board (OSB)); demand for wood and final goods; and policies that affect incentives to produce public goods and tax policy.

First and foremost, benefits will depend on land quality and the producer's location. Joss et al. (2008) conducted studies of land suitability for the growth of hybrid poplar trees in Saskatchewan, and focused on the northern agricultural zone near the forest fringe, mapping growing potential by moisture deficit and soil type. The northern area has the most potential for wood production in agroforestry or afforestation, given its close proximity to the existing forest infrastructure and the increased potential for farmers to sell wood products. It is also the region in which hybrid poplars can suitably be used in shelterbelts. In the northern agricultural zone, a producer who grows trees for dimension lumber, OSB or pulpwood will have an advantage due to better moisture balance and lower transportation costs, given that transportation is a significant factor in determining long-term profitability. However, such producers will face competition from trees grown on crown lands, and will operate in commodity markets that typically generate lower revenues[8].

Alternatively, producers who target niche markets, such as wood used for furniture or specialty products, will have an advantage since proximity to existing forest infrastructure may not be as important, and because wood used in niche markets is often more valuable. The AAFC-AESB Agroforestry Development Centre has conducted many studies with respect to growing potential for different tree and shrub species and has made selections of superior genetic materials for some of the species that will be valuable when determining species and potential crop profitability.

Also important is the location of final goods markets. The forest industry currently produces commodity products destined for the US or international markets. The Prairie Provinces are disadvantaged in two ways. First, the local populations are relatively small and therefore do not serve as large final goods markets. Second, transportation routes to the US and to either coast are lacking thus adding to the costs of production and reducing competitiveness.

Many successful agroforestry systems rely on joint production of more than one crop such as in the case of intercropping and silvopasture, which would rely on the economic model of joint production. In the case of silvopasture, trees can enhance livestock production by providing shade to reduce the stress caused by heat and wind. Trees can then be harvested at maturity on a rotational basis to add revenue to the producer's operation. In

---

[8] This aspect is further discussed in the subsequent paragraphs.

addition, animal waste can be liquefied and used to fertilize crops. Finally, there could be environmental benefits in the way of improved water quality and riparian habitat that add value.

Producers must be cautious however, to avoid negative side effects. One of the greatest costs of silvopastoral operations is protecting trees from livestock. This includes not only the period for which the woody crop is established, but subsequently from browsing and rubbing, the latter of which will cause trees to have deformities, thus reducing market values. Production efficiency will be important and producers must pay attention to how both products will be produced and managed – as complements or substitutes.

Ultimately, long-term profitability in hybrid poplar plantations will depend on production decisions, such as crop selection and rotation, and an understanding of markets, long-term financial analysis, risk analysis, and valuation. The economics of afforestation are similar to that of forestry and must consider operating costs from plantation establishment to fertilization, irrigation (if applied) and pruning, for example. There might also be a significant opportunity costs given the time to maturity of the crop and the absence of a revenue stream in the interim. Farmers engaged in the production of annual crops receive revenue annually, whereas revenue for woody crops can take decades. The cost of time must be factored in. Individual producers who establish hybrid poplar plantations must begin with the end in mind, while understanding the constraints of production.

Additionally, risk must be factored into the analysis and can take many forms. There are biological risks that could reduce the value of a crop including loss from insects and disease, browsing by ungulates, drought, hail or flood, or unforeseen events, such as fire and blow down. By understanding these risks, producers can factor them into the long-term economic assessment to have a better idea of expected profits. Similarly, there are market risks resulting from changes in demand, transportation and fuel costs, new forms of competition from other producers and new products. Finally, producers are also affected by political risks pertaining to agricultural policy in areas related to taxation and carbon sequestration, development of emerging markets, the manner in which they are regulated, fluctuating prices and how property rights are defined.

Evaluating the private benefits of an agroforestry system, such as plantations, requires financial analysis that determines the net present value[9] of

---

[9] This is not to suggest that all types of agroforestry must go through this type of evaluation. For example, shelterbelts are planted for reasons related to private and social benefits. Many

all decisions, and that selects the appropriate discount rate to reflect the risks involved and the value of time (Filius, 1982). In such an analysis, the private producer will be concerned with financial benefits, although that could include the value of public goods produced. This latter point is particularly important if government policy is formulated to compensate private interests for producing public goods.

Private benefits from agroforestry can be generated in two distinct pathways: (1) direct benefits from the production and sale of the agroforestry products, and (2) agroforestry can generate indirect benefits to land owner through improvements in other farm level enterprises. Both of these types of benefits are described in this section.

Different forms of agroforestry have the potential to provide benefits to the land owner. Trees and their products can be sold or consumed for home-use, as most of these are commercial products[10]. Unfortunately for the Canadian prairies, the only forms of agroforestry for which benefits have been studied are tree farming and shelterbelts.

Lindenbach (2000) considered agroforestry as an agricultural crop competing with other crops and assessed its economic feasibility comparing it with three rotations: A tweaked (adjusted) rotation[11], a grass rotation, and a hybrid poplar rotation. The study concluded that hybrid poplar under current prices and costs is not economically viable. The high cost of harvesting was the main reason attributed for the poor economics. This study also undertook simulations under alternative conditions. When producers were paid $20 per tonne for carbon sequestration, hybrid poplar returns became positive.

Dore et al. (2000) examined planting trees (mainly hybrid poplar) on marginal agricultural lands in northern Saskatchewan. This study concluded that conversion to trees had both economic and ecological advantages, thereby creating several positive externalities[12]. In a subsequent study, Dore et al. (2001) applied stochastic dominance to test the dynamics of converting land from agricultural production to afforestation. The analysis indicated that afforestation of these lands is socially desirable particularly if carbon sequestration benefits are taken into account.

---

of the social goods being non-market goods in nature, cannot be evaluated in a benefit-cost analysis type framework.

[10] For other farm level benefits see Josiah (2000).

[11] A tweaked rotation included a typical crop mix in the northern part of Saskatchewan. It included spring wheat, barley, oats, peas, flax, and canola as the typical crops grown.

[12] Some of these externalities are presented as a part of the discussion on social benefits in the next section.

Johnston et al. (2001) suggested that acceptance of hybrid poplar plantations could be improved if changes in taxation of operations, property assessment, and recognition of the value of carbon sequestration are taken into account through formulation of proper policies. In a further study, Johnston et al. (2002) also suggested that distribution of revenues from timber sales to the landowners earlier in the life of the plantation might lead to better acceptance of plantations and other types of agroforestry.

Protection of crops by field shelterbelts clearly results in yield increases and, for some crops, improved crop quality. Some cropping situations such as potatoes on sandy soils, small-seeded vegetables in rotor-tilled beds or sensitive fruit and vegetable crops, return the greatest shelterbelt benefits (Kort, 1988).

Another type of private benefit is related to land reclamation. This is the lowering of the productivity of land through processes such as soil erosion, loss of soil fertility, and increases in soil salinity. Trees can be effective tools for land restoration, particularly for salinity control, although by no means are they the only way to solve this problem (Cacho, 1999). Notwithstanding, the beneficial effects of trees on land productivity from reduction of salinity translates into higher land value.

4.2. PUBLIC BENEFITS

There is a long list of public goods produced through agroforestry systems from which society benefits. That is to say, society receives such goods, yet does not pay for their benefit and use. Examples of such benefits might include, among others, enhanced soil and water quality, soil and water conservation, improvements in air quality (particularly around swine operations), greenhouse gas sequestration, enhanced wildlife and species conservation, and improved biodiversity (Alavalapati and Mercer, 2004). If such public benefits are valued by society, there is a strong rationale to develop policies that provide incentives for the private sector to further adopt agroforestry practices.

Most benefits of agroforestry to society are typically generated though various pathways related to ecosystem functions (Kulshreshtha and Kort, 2009). These include: **Soil** (reduced soil erosion; shoreline protection); **Air** [reduced odours from animal production sites; reduced pesticide drift (also affecting water quality indirectly); reduced greenhouse gas accumulation in the atmosphere]; **Water** (increased water quality through filtering function;

floodplain management; wastewater management); **Biota** (wildlife habitat enhancement; wildlife based recreation; increased biodiversity). In addition, agroforestry could also benefit society through (1) energy conservation; (2) aesthetic and related amenities; (3) improved farm level economic efficiency; (4) transportation infrastructure and traffic related impacts; (5) improved health impacts; and overall improvement of the quality of life. Some of these pathways, as shown in Table 3, generate both private and public benefits.

Table 3. Taxonomy of Benefits from Agroforestry Operations

Category / Pathways	Type of Activity	Private Benefits	Social (Public) Benefits	Joint Private and Public Benefits
Commercial	Sale of Agroforestry products	X		
	Improved revenues from crop and livestock enterprises with agroforestry	X		
Potentially Commercial	Carbon sequestration	X		
Soils	Soil Quality – Reduced soil erosion			X
	Improved air quality – Reduced particle drift		X	
Water	Water quality improvement			X
	Floodplain management		X	
	Wastewater management		X	
Air	Improved air quality – Reduced odour			X
	Reduced Greenhouse Gas Emissions		X	
Biota	Increased biodiversity		X	
	Wildlife habitats -- Recreation		X	
Others	Energy conservation			X
	Improved aesthetics			X
	Shoreline protection		X	
	Transportation safety		X	

4.2.1. Soil Resources Pathway

Certain types of shelterbelts plantings can improve soil quality. Field and riparian shelterbelts are two such examples. Field shelterbelts reduce the damage from wind erosion by reducing the effect of wind on crops, particularly during the early stages of establishment. Under rolling topography, these can also reduce water erosion, which is a major problem in southwest Saskatchewan and southeast Alberta (Smith and Hoppe, 2000, p. 100). Riparian area shelterbelts help in the restoration and stabilization of shoreline and thus, help reduce water erosion. Each of these biophysical changes is socially relevant, leading to benefits either to landowners who plant these shelterbelts or to society at large.

Soil erosion creates problems for the producers as well as to other members of society. Therefore, the traditional measure of the amount of soil displaced presents only a partial picture. The real measure of severity of wind erosion is the cost of the damage it does (Huszar and Piper, 1986). Wind erosion costs are incurred both on and off the farm. On-farm erosion costs include decreased soil productivity through loss of soil organic matter in the topsoil. This results in lower yields to producers, and therefore, is a private loss. Shelterbelts can reduce the rate of soil erosion and thus reduce losses to landowner.

Off-site costs of soil erosion are incurred by individuals, local governments, industry, and regional and/or provincial governments. Some examples of these costs are for: non-point source pollution; increased sedimentation in water channels (drainage and watercourses) and reservoirs; adverse effects on social activities, such as sport fishing or camping; water treatment; and health related incidents. The off-farm costs of soil erosion occur as sediment and other erosion-related contaminants enter streams and lakes (Dickson and Fox, 1989).

4.2.2. Water Resources Pathway

Deterioration of surface water bodies is primarily through run-off from crop production and livestock operations. Plants and trees in riparian areas can filter sediment from agricultural land runoff; their stems slow and disperse flow of surface runoff, and promote settling of sediments. In addition, their roots stabilize the trapped sediment and hold riparian soil in place. Similarly for agricultural (livestock) operations, trees can filter nutrients, pesticides, and animal waste. Thus, shelterbelts and other types of agroforestry can improve off-site surface and groundwater quality.

The use of agroforestry practices to reduce non-point source pollution from monocropping areas, by using buffer strips of woody vegetation and integrating trees into cropping systems, appears to have excellent potential (Brooks, Gregersen and Ffolliot, 1994).

Over the last century, natural floodplains have been altered. Land clearing has resulted in the loss of woody vegetation in areas adjacent to streams and rivers. These changes have increased the risk of flood damages in various parts of the world. Planting of waterbreaks, through re-establishing trees, is a recommended practice in the US midwest (Wallace et al., 2000). The use of waterbreaks during flooding can trap debris, reduce sand deposition, and reduce damage to roads and ditches. Tree trunks will reduce floodwater velocity and erosive power and block stream debris (Dosskey, Schultz and Isenhart, 2003).

Hybrid poplars are fast-growing trees and are thus well suited to use for agricultural, community and industrial wastewater (Kuhn and Nuss, 2000). In this manner, trees are an alternative to more expensive wastewater treatment systems. Projects involving the use of alley-cropping to treat municipal sludge have been reported in the US, and it is suggested that similar systems can be developed for the disposal of manure from the large livestock production facilities (Schultz et al., 1995). Several projects are underway in Alberta investigating the use of municipal wastewater for irrigating fast growing willow plantations.

4.2.3. Air Resources Pathway

Planting of trees has been suggested to improve air quality by reducing odours from confined livestock operations (hog barns and feedlots). According to Tyndall and Colletti (2000), livestock production system related odours can be dispersed with the help of shelterbelts.

Trees can have a significant impact on the ambient air quality of the region. Forested lands can clean air of microparticles of all sizes twenty times better than barren land (Griffith, 2003). Since particles smaller than 10 microns can become permanently lodged in the smallest sections of the lungs, their reduction could have some health-related benefits for society. Some of the significant airborne sediments are pesticides. Besides affecting health of individuals, their major effect is on water quality and through that on the health-related ailments of humans.

In addition to sequestering atmospheric carbon in biomass, shelterbelts and other agroforestry systems can also reduce greenhouse gases directly[13], as well as through reductions in emissions from crop production. Land taken out

---

[13]   This aspect of the agroforestry is discussed in Section 4.2.5.

of production would reduce the amount of nitrogen fertilizer required on a field, thus reducing GHG emissions (Thomsen Corporation, 1999). Deep-rooted trees in shelterbelts can intercept nitrogen leaching into the soil and take up a proportion of it. Thomsen Corporation (1999), for example, estimated this uptake to be approximately 18 kg ha^{-1} per year. In addition, the leaf litter from shelterbelts can contribute 5 kg ha^{-1} yr^{-1} of N, thus removing the required amount of inorganic N fertilizer.[14]

The total amount of reduction in greenhouse gases from adding shelterbelts to the farm landscape was reported by Kulshreshtha and Junkins (2001) to be 2,094 kilotonnes in carbon dioxide equivalent (CO$_{2e}$). This estimate is based on converting 1% of the cropped area in the three Prairie Provinces to shelterbelts.

There has been research in the last five years in the Prairie Provinces into biomass production for bioenergy production, primarily through the use of willows, building on past research in Europe (mainly Sweden), Syracuse, NY, and in Quebec. Woody biomass can replace fossil fuels (coal, gas, oil) in producing heat, electricity and fuels, such as ethanol or bio-gas. As compared to the sequestration of carbon in shelterbelts or plantations, renewable and sustainable biomass production can reduce net carbon dioxide emissions through fuel switching.

4.2.4. Biodiversity Related Pathways

Agroforestry can produce a positive impact on society through a number of biodiversity related pathways. Three such pathways are: (1) an increase in the number of species of plants and wildlife in areas containing shelterbelts or plantations as well as in the surrounding areas; (2) society may develop some benefits from non-consumptive use of wildlife (such as bird watching or wildlife viewing); and (3) members of society may use wildlife for consumptive (hunting) purposes.

Shelterbelts and other agroforestry systems benefit wildlife habitat, since they supply perennial vegetation, diversity of cover and food for wildlife (Dosskey, Schultz and Isenhart, 2003). They provide "green corridors" for wildlife to travel from one habitat to another. The abundance of small mammals and herpetofauna[15] increased with vegetation structure complexity.

---

[14] One should note that this only works if the leaf litter is spread over the field.

[15] Definition of a herpetofauna is Amphibian and Reptile fauna.

Small mammal diversity was higher in herbaceous and wooded riparian strips, whereas the herpetofaunal community was more diverse in shrubby strips (Maisonneuve and Rioux, 2001, p. 165).

Great Plains riparian woodlands have been reported to contain bird communities that are as much as seven times as rich as those in the surrounding plains (Chranowski and Antonowitsch, 1995). In this study, a positive correlation between Habitat Suitability Index for various types of shelterbelts and bird species richness with the presence of trees was reported.

The effect of agroforestry on soil biota is not as frequently reported. Atlavinyte (1964) reported a loss of soil organisms under conventional agriculture compared to more sustainable production systems (including agroforestry). Price and Gordon (1999) found that earthworm densities were greater in intercropping systems than conventionally cropped fields in southern Ontario. In addition, measures that enhance soil organic matter through litter dynamics (such as shelterbelts and other agroforestry operations) should favour proliferation of soil biota.

4.2.5. Carbon Sequestration

Carbon sequestration has been determined for Canadian prairie shelterbelts (Kort and Turnock 1999) and there are data for hybrid poplar plantations. Kort and Turnock (1999) measured aboveground biomass in mature shelterbelts of the most common shelterbelt species, including deciduous and coniferous tree species and shrubs. Regional differences in moisture deficit corresponded with differences in tree and shrub growth rates, which increased from the drier Brown Soil Zone, to the less dry Dark Brown Soil Zone and finally, to the Black Soil Zone. Average shelterbelt carbon content aboveground was 32 t C km^{-1} for green ash (average age: 53 y), 41 t C km^{-1} for white spruce (average age: 54 y), 105 t C km^{-1} for hybrid poplar (average age: 33 y) and 26 t C km^{-1} for caragana (average age: 49 y). Other species were also measured. The data for hybrid poplar plantations include experimental plantings in which various poplar clones are being tested, and data taken from poplar shelterbelts. Most of the plantations have not reached full rotation, but data from 12-year old stands in the Aspen Parkland Ecoregion and other literature suggest that stands of hybrid poplar may yield 40-60 tonnes of aboveground carbon per hectare (t C ha^{-1}) at age 20 (Peterson et al. 1999). This suggests an annual rate of sequestration of 2-3 t C $ha^{-1}y^{-1}$ as compared to rates of sequestration in the region's boreal forest of 0.4 – 0.6 t C ha^{-1}y^{-1} (Peterson et al. 1999).

Results for other agroforestry systems in the Prairies are not available. However, as a proxy, in southern Ontario, Peichl et al. (2006) investigated total carbon sequestration and annual carbon fluxes in two intercropping systems and a conventional cropping system. In the intercropping systems, 13-year-old hybrid poplar (*Populus deltoides* X *Populus nigra* clone DN-177) and Norway spruce (*Picea abies* L.) were intercropped with barley (*Hordeum vulgare* L. cv. OAC Kippen). In the conventional agricultural system, barley was grown as a sole crop. Total C pools (including an assumed barley C pool of 3.4 and 2.9 t C ha^{-1} within the sole cropping and the intercropping systems respectively) were 96.5, 75.3, and 68.5 t C ha^{-1} within poplar, spruce intercropping and in barley sole cropping systems, respectively. Estimated net C fluxes for the poplar and spruce intercropping systems and for the barley sole cropping system in 2002 were +13.2, +1.1, and -2.9 t C ha^{-1} y^{-1}, respectively. While growing conditions are somewhat better in Ontario than in the prairie region, these results suggest that intercropping systems have a greater potential in reducing the atmospheric carbon dioxide concentration compared to sole cropping systems.

A challenge for using agroforestry systems in carbon credit applications is the change in carbon stocks that occurs with harvesting or, for shelterbelts that may never be harvested, the changes that occur after maturity when they may decline in vigour or if they are replaced or removed. Plantations of fast-growing species would typically be harvested every 15-20 years, and intercropping systems are more complicated, with both trees and crops harvested at varying time intervals. For plantations, a relatively straightforward solution is to plant 1/R of the total plantation area per year, where R is the rotation age (i.e. the age at which the trees are harvested). The last fraction of the area is planted in year R-1 and the first fraction planted will be ready to harvest in year R. To illustrate this concept, a fast-growing variety of hybrid poplar harvested at 20 years is assumed. The planting rate adopted is that currently being used by Al-Pac, a forest products company in northern Alberta that is planting 1,200 ha y^{-1} through leasing arrangements with local land owners. This yields a total area of 24,000 ha over 20 years. When the system reaches a steady state at 20 years (i.e. no additional area is being planted or harvested), approximately 660,000 t C is sequestered in perpetuity as long as each stand is replanted after harvest. Figure 2 illustrates the time course of carbon sequestration on the entire 24,000 ha landscape. Note that carbon is added to the system each year until year 20, after which no additional carbon is sequestered. Therefore, any carbon credit value lapses at year 20, and the carbon sequestered to that point must be maintained as dictated by the rules of the carbon credit program, typically for 20-30 years.

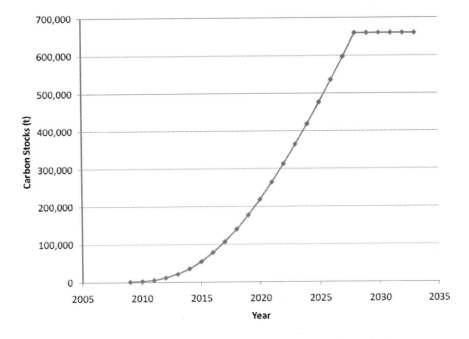

Figure 2. Time course of carbon sequestration based on hybrid poplar plantations.

The trading of credits will occur in the private sector after being certified by the Offset System and transferred into the proponent's account. Various trading platforms similar to Chicago Climate Exchange (CCX) are in development but none is active at this time.

4.2.6. Other Benefits

In addition to the above set of benefits to society and producers, agroforestry and in particular shelterbelts could also provide other benefits. One of these benefits may pertain to improved aesthetics of the property. Although shelterbelts may result in a greater resale value to the house and property, farm properties not close to urban areas are generally purchased by other farm families and the effect of shelterbelts on property resale value may not capture their true value. According to Kulshreshtha and Knopf (2004), a good farmyard shelterbelt may not increase the property value, but makes it much easier to sell. Reduction in traffic accidents due to trees' ability to reduce snow accumulation on roads, and energy conservation benefits to the society, are some other examples of such benefits (Kulshreshtha et al. 2005).

Estimation of public benefits of various types of agroforestry operations has not been done in a systematic manner. Public benefits from shelterbelts over the 1981-2001 period for the shelterbelt plants distributed by the Agroforestry Development Centre of Agriculture and Agri-Food Canada[16] were reported to be $140 million (Table 4).

Table 4. Summary of estimated external benefits from shelterbelts through AAFC Shelterbelt Centre activities, 1981-2001, using a 5% discount rate

Pathways	Biophysical Impact	Level of Benefits from Public Goods (Mill. $)	Unquantified Benefits
Soil	Reduced soil erosion	$14.8	Shoreline stabilization
Air	Improved air quality (Non-odour related)	$3.7	Odour Reduction
	Reduced Greenhouse Gas Emissions through Carbon sequestration	$56.0	Reduced pesticide drift
	Reduced Greenhouse Gas Emissions through reduced cropped area	$16.6	
Water	Improved water quality	$1.2	Wastewater management
Biota	Biodiversity	$4.7	
	Consumptive wildlife based recreation	$39.1	
	Bird Watching	$3.7	
Other	Energy Conservation based GHG emissions reduction	$0.2	Aesthetics and Property values
			Transportation
			Health Impacts
Total Benefits		$140.0	

---

[16] More details on the method of estimation of these benefits are provided in Kulshreshtha and Knopf (2004).

5. MAJOR ISSUES IN AGROFORESTRY DEVELOPMENT ON THE PRAIRIES

5.1. PRODUCERS' PERCEPTIONS OF BARRIERS TO AGROFORESTRY

Lack of adoption of agroforestry may have been a result of many factors, some of them may have been rooted in technical, economic, financial, or environmental concerns. A partial list of barriers identified during a focus group survey of prairie producers is presented in Table 5 (Marchand and Masse 2008). Although no weighting of each concern was provided, the range of such factors is wide. Technical barriers include lack of machinery for various agroforestry operations, expertise requirements, and low probability of a successful operation. Economic / financial reasons for lack of adoption were related to lower profitability, and lack of revenue until harvest.

5.2. SOCIO-ECONOMIC CHARACTERISTICS OF PRODUCERS

The prairie region has a high proportion of farm operators that are older. Table 6 shows a distribution of age of farm operators in the three Prairie Provinces. In 2006 some 44% of all farm operators in the region were of 55 years or older. The proportion on a provincial basis was very similar ranging from 41% in Manitoba to 45% in Alberta. This older age distribution makes many farm operators less interested in long term investments such as that related to the development of commercial agroforestry systems. Along with this older age distribution is the mindset that many of these farmers have spent a lifetime removing trees from their landbase to convert to pasture or agricultural crops thus making it difficult to convince them to plant trees on their farms.

Table 5. Major Factors Cited in the Lack of Adoption of Block Plantation of Hybrid Poplar in the Prairie

Category of Concerns	Description of Concerns
Technical	• Machinery required for site preparation and maintenance • Site preparation methods still under development • Planting not mechanized • Lack of knowledge transfer and technical training • Maintenance work required during the initial years • Requires a high level of expertise and experience in enterprises wishing to produce hybrid poplar • Few specialists to evaluate soil quality and fertilization needs • Probability of planting success low or uncertain • Stump removal necessary after harvesting • Risks associated with large-scale monoculture • Possibility that hybrid poplars may obstruct or damage agricultural drains
Economic / Financial	• Substantial establishment and maintenance costs • Establishment of hybrid poplar not subsidized • Period of 15 to 20 years before planting will produce any income (long-term investment) • Property assessment goes up if land is afforested • Tax collected at harvest discourages investment for the establishment of plantations • Existing markets (pulp, panelboard, lumber, peeling) not obvious prospects • No carbon sequestration market • Less profitable than agriculture in the case of good land
Environmental	• Risk of a decline in biodiversity, as deciduous forests are usually mixed • Risk that mechanical weeding may cause soil erosion • Less carbon sequestration than in the case of spruce

Source: Adapted from data presented by Marchand and Masse (2008).

Table 6. Age Distribution of Prairie Farm Operators, 2006

Age category	Manitoba	Saskatchewan	Alberta	Prairies	Prairies % of Total
<35 years	1,100	2,815	2,045	5,960	8.6%
35 - 54 Years	5,905	13,700	13,335	32,940	47.3%
> 55 years	4,785	13,235	12,665	30,685	44.1%
Total	11,790	29,750	28,045	69,585	100.0%

5.3. CLIMATE CHANGE AND AGROFORESTRY

Carbon dioxide (CO_2) and other greenhouse gasses (GHG) in the earth's atmosphere have been increasing since the industrial revolution in the 1850s and are likely the cause of recent global warming. The latest climate change assessment from the Intergovernmental Panel on Climate Change (IPCC) concluded:

> "Most of the observed increase in global average temperatures since the mid-20th century is *very likely* [> 90% probability of occurrence] due to the observed increase in anthropogenic GHG concentrations." (IPCC 2007).

CO_2 is a long-lived gas in the atmosphere and therefore a certain amount of global warming would occur even if current emissions were stopped immediately (Ramanathan and Feng, 2008). Part of the solution is to increase withdrawal of CO_2 from the atmosphere. Biological processes such as photosynthesis and soil organic matter formation remove CO_2 from the atmosphere and sequester it in vegetation and soil; this is known as a biological carbon sink. Sustainable environmental management practices, including agroforestry systems, enhance the functioning and maintenance of carbon sinks and therefore represent one of the most important values in appropriate forest and agricultural land use practices.

The Canadian Government ratified the Kyoto Protocol (KP) in February 2005. The KP recognizes afforestation and other agricultural land use activities as an important carbon sequestration option that countries are required to include in their GHG inventories. In addition, the Canadian domestic Carbon Offset System, currently in the final design stages, will include agricultural and forest-related carbon sinks. This will provide the basis for a carbon credit trading system in which landowners can realize financial benefits from creating and maintaining carbon sinks. Therefore agroforestry and afforestation have a large potential for contributing to GHG reductions through sequestration or bioenergy production in the prairie region of Canada.

Greenhouse gas emissions in the prairie provinces (Alberta, Saskatchewan and Manitoba) in 2006 were 327 million tonnes of carbon dioxide equivalent (Mt CO_2e), while Canada's emissions were 721 Mt CO_2e (Environment Canada 2008). Alberta emitted 72% of the regional total, with Saskatchewan emitting 22% and Manitoba 6%. Manitoba's emissions are low due to a wealth of hydropower resources, while Saskatchewan and Alberta rely mainly on coal-fired power sources. The relatively higher emissions from Alberta and Saskatchewan are also due to an active oil and gas sector, with oil sands development prominent in Alberta. Figure 3 shows the proportional contributions from various sectors. Regionally, energy production accounts for 82% of CO_2 emissions, agriculture 12% and the remainder is from other industrial processes and waste.

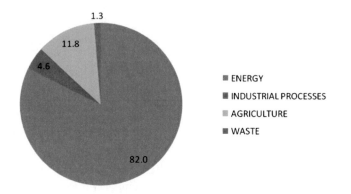

Figure 3. Percent emissions for various sectors in the Canadian prairies (Alberta, Saskatchewan and Manitoba). Data from Environment Canada (2008).

The recent Canadian National Climate Change Assessment (Lemmen et al. 2008) summarizes the impacts of climate change on the prairie region as follows:

- Increases in water scarcity represent the most serious climate risk in the Prairie Provinces.
  Suitability for agroforestry is generally highest along the northern edge of the current agricultural zone in the prairie region. For

example, Joss et al. (2008) found that approximately 150,000 km^2 is suitable for hybrid poplar in this region. Climate change projections for this area include warming temperatures and little increase in precipitation (Lemmen et al. 2008). This will result in increased rates of evapotranspiration and decreased soil moisture availability, affecting the potential for agroforestry. Careful site selection for agroforestry establishment will be required in order to focus efforts on areas that will maintain productivity in the future. In addition, selection of trees and crops for drought tolerance may be required, or species and clones currently considered suitable for drier southern regions may be used further north.

- Ecosystems will be impacted by shifts in bioclimate, changes in fire and insect disturbances, stressed aquatic habitats and the introduction of non-native species, with implications for livelihoods and economies dependent on ecological services.

Although current disease and insect pests in shelterbelt species are well known, the possible effects of climate change are unknown. New threats by non-native pests like the Emerald Ash Borer may be come more of a threat under some climate change scenarios. The lack of experience with hybrid poplar plantations in the prairies makes predicting the impacts of pests and diseases difficult even under current climate. In the future, the expectation is that warming temperatures and drought stress may encourage existing pests to expand their range, and also to allow pests from more southerly locations to move into the region. Agroforestry development will need to focus on establishing and maintaining healthy crops and trees and to explore opportunities for breeding or selecting pest-resistant species, clones or varieties. The risk of fire is also expected to increase in the future, with some estimates projecting a 2-3 times increase in area burned for the western Canadian boreal forest by late 21st century (Balshi et al. 2009). However, as agroforestry practices occur in developed agricultural areas with generally good road access, reducing the likelihood of fires relative to the natural boreal forest.

- Communities dependent on agriculture and forestry are highly sensitive to climate variability and extremes. Drought, which can have associated economic impacts of billions of dollars, wildfire and severe floods are projected to occur more frequently in the future.

Careful economic analysis will be required to understand the impacts of climate change on communities that may become dependent on afforestation for its livelihood.

- Adaptive capacity, though high, is unevenly distributed, resulting in differing levels of vulnerability within the region.

  Small rural and aboriginal communities may be more vulnerable to the effects of climate change. This factor should be taken into account when planning for agroforestry development.

Conversely, climate change may increase forest productivity and fibre yield given increasing carbon stocks (Johnston and Williamson 2005; Sohngen and Sedjo 2005). If increases in yield outweigh negative effects (fire, insects and disease) total regional carbon stocks will increase. In this case, the increase in forest productivity and resulting supply will depress global prices resulting in decreasing producers' surplus and increasing consumers' surplus (Sohngen and Sedjo, 2005).

Increasing disturbance may also provide opportunities for producers of tree plantations. Investing in species that are drought resistant and less susceptible to insect infestation and disease will increase the probability of long-term operational success. Similarly, increased fire severity in the natural forest may provide farmers with production advantages where private crops are less susceptible to conflagration.

Whereas climate change factors that affect productivity and disturbance regimes cannot be altered, attention to how such changes will affect demand, supply and prices in the long run will help producers to position themselves so that they can mitigate effects. Research and development to identify final goods and international markets, global supply and biophysical properties of species that are well suited to conditions in central and northern prairie region will help producers to be successful in this regard.

5.4. WATER ISSUES RELATED TO CLIMATE AND ECONOMICS

Change in hydrological cycles as a result of climate change will affect exiting forest stocks on public and private lands (Farness and Hesseln, 2006). Such changes will affect the condition of forests and woody crops through increased insect infestation and disease prevalence (McKinnon and Webber,

2005). Similarly, wildfire and increasing insect-fire interactions is likely to reduce the total supply of fibre, primarily in the boreal forest (Volney and Hirsch, 2005). Under each scenario, this may decrease total supply of wood fibre derived from both public and private land.

Shelterbelt tree and shrub species range from drought-hardy shrubs to moisture-loving species like poplar and willow. Increased drought due to climate change may reduce the species options for landowners or impair the performance of the species that landowners plant.

5.5. BIOENERGY NEEDS

Bioenergy in the last few years has gained momentum in the prairies, and may have implications for agroforestry practices and designs. With the ever increasing focus on the environment and going 'green' by the general public, climate change and rising oil prices, the use of renewable energy has come to the forefront again after last appearing in the 1970s. Growing biomass energy crops such as willow, switchgrass or other fast-growing species may be more appealing to farmers as the rotations are considerably shorter than growing hybrid poplar for wood fibre for saw or pulp mills. Rotations for willow are generally three years in length; however, initial start up costs are extremely high ($6,000 to $10,000 ha^{-1}) during this early stage of development. The majority of the cost is the planting material as $\sim$ 15,000 stems are planted ha^{-1} and planting stock costs or incentives will be needed to establish these systems. Cuttings of hybrid poplar from various growers across the Prairies can be purchased from $0.20 to 0.35 depending on quality and willow cuttings in the early stages of development will probably fall in this range until production is ramped up. In Europe, the establishment costs are as low as $1,500 ha^{-1} with seedling costs at < $0.10. However, once established these systems can function for over 20 years without having to be replanted. As willow or other biomass crops are relatively new in the prairies, little is known about what species or clones are suitable for the soil and climatic conditions of the prairies and much research is needed to provide the answers to basic agronomy questions. The amount of planting material needed however, is significant compared to hybrid poplar plantations (1,500 stems ha^{-1}) and a process for ramping up the supply of suitable material for planting will need to be developed – a problem which also effectively stalled the large scale plantings in the late 1800s. The development of markets has not occurred as yet on the prairies although there is a lot of interest in producing ethanol, biomass for electricity or other products.

There is also an option of producing woody biomass in multi-functional agroforestry systems such as shelterbelts or riparian buffers. Although fast growing species such as poplar and willow would be suitable in some of such systems, they would be largely justified because of their other environmental benefits. While the harvest and sale of biomass for bio-energy would be an added incentive for the development of such systems, maximizing growth would not be the only criterion for them, so that a diverse palette of tree and shrub species could be used. From an economic perspective, these types of agroforestry plantings that have important environmental benefits can often receive cost-sharing by government of non-government organizations and occur on the margins of fields or riparian zones where they do not require the conversion to trees of valuable cropland."

6. POLICIES RELATED TO AGROFORESTRY DEVELOPMENT

Agroforestry practices that include shelterbelts, wildlife habitat, riparian and other environmental plantings have been continuously supported on the prairies through the Government of Canada, as discussed in the "History" section, through the supply of adapted tree material and associated recommendations, scientific investigation and administration, delivered through the AAFC-AESB Shelterbelt Centre. As well, the Government of Canada has directly delivered cost-shared or incentive programs and partnered with provinces in such programs. The Prairie Provinces have also directly supported agroforestry through initiatives such as the former shelterbelt program in Alberta and the Manitoba Agro-Woodlot Program. The support for shelterbelts has been relatively constant over the years, although special emphasis was placed on field shelterbelts for erosion control in the 1930's, the 1960's and the 1980's. Meanwhile, support for other agroforestry practices have changed over time, from encouraging fuelwood plantations in the early years, to increased emphasis on special environmental plantings in the last 25 years, including wildlife habitat plantings, riparian buffers and shelterbelts as a carbon sink and to special agroforestry designs which also provide economic products (wood and biomass, fruit or maple syrup). The main policies which the federal and provincial governments have adopted to encourage these practices have been to provide trees and shrubs, to provide field staff, infrastructure support and expertise and to share the costs of agroforestry establishment. Cost sharing has taken the form of in-kind support (e.g. specialized equipment, plastic weed barrier) or direct payments to landowners for an eligible percentage of the costs of establishment.

Agroforestry was mentioned in Saskatchewan's State of the Environment Report (1997). An Advisory Council on Agroforestry was also formed to provide the provincial government with advice on how to proceed in the future. The activities included woodlot management, shelterbelts, wildlife plantations, maple syrup production, berry harvesting and processing, Christmas tree production and grazing systems on forested lands. To encourage the adoption of sustainable farming practices, funding was committed to promote the stewardship of soil, water, air and biological resources to produce food and fibre into the future with one of the five priority areas to be the research, development and promotion of sustainable agroforestry for economic diversification opportunities.

In 1997 and 1998, under funding from the Canada-Saskatchewan Innovation Fund, the AAFC Shelterbelt Centre, in partnership with wood-processing industries and private landowners, established six hybrid poplar test plantations of 7 hectares each that included replicated clonal and spacing trials. These early plantations spurred broader interest and discussion. In 2001, the Federal and Saskatchewan governments announced the creation of the Saskatchewan Forest Centre (known till 2009 as ForestFirst) which, among other responsibilities, had a mandate to develop fast-growing tree plantations in the province. The Centre's agroforestry unit has played an important role in the province in promoting and developing various demonstrations of tree establishment practices and transferring related research to farmers through extension meetings, courses and print. This agency has been crucial in providing a grass roots effort to develop plantation-based agroforestry systems. In 2003, the Saskatchewan Ministry of Agriculture through the Agri-Food Innovation Fund created the AFIF Chair in Agroforestry and Afforestation in the College of Agriculture and Bioresources at the University of Saskatchewan with a mandate to conduct research and develop an afforestation industry in the province of Saskatchewan.

The concept of poplar farming started gaining some momentum in the next few years which was bolstered by the Hon. Lorne Calvert's 2005 throne speech that stated "Agroforestry, including the growing of trees as crops, is underway. My government has a bold vision for the future of this industry. Over the next twenty years, the goal will be to transform ten per cent of the arable land of Saskatchewan to agroforestry, creating another sustainable industry in our province."[17]

---

[17] For details see Government of Saskatchewan (2005).

This exciting news certainly did encourage the agroforestry community in the province and it was re-iterated in the throne speech in 2006. However, in the fall of 2007, governments changed and the vision of trees being planted began to fade and agroforestry has not been mentioned in the throne speech again. Agroforestry is a long-term commitment to a crop as even fast-growing hybrid poplar plantations require from 3 to 20 years to grow before harvesting. Governments and farmers tend to think of one year cycles and this mind set is important to change if using trees as an agricultural crop is going to play a role in resource development in the prairies in light of developing products, markets and research funding.

The Forest Service Branch of the Ministry of the Environment implemented a policy to encourage planting trees on agricultural land whereby the forest industry would have to obtain 20% of its wood supply from private land. Landowners could utilize their native forests for this wood supply, although companies would not assume any responsibility for the sustainable management of these lands. Another option is the development of hybrid poplar plantations by farmers to supply mills with wood fibre. Any operation that was located in close proximity to the mills would have an advantage in reduced transportation costs. Additionally, farmers would increase non-market benefits in the way of increased biodiversity and market value through possible carbon credits. Initial costs of plantation establishment however, are high and would deter farmers from entering into such a venture unless incentives were provided or other arrangements (i.e. lease arrangement via Al-Pac model) were instituted.

Agriculture and Agri-Food Canada's Agri-Environment Services Branch (AESB) supports agroforestry activities throughout Canada through Federal-Provincial partnerships to deliver cost-share programs to encourage the adoption of agroforestry and other Beneficial Management Practices, and by providing technical expertise and local administrative support through AESB regional staff. Agroforestry priorities vary from region to region and depend on such local factors as land use practices, land value, risks to water bodies, biodiversity, etc.

7. Policy Needs for Fostering Agroforestry Development

The success of the emerging poplar farming industry in the Prairie Provinces will depend greatly on how government policy is developed and the incentives for producers to convert from annual crops to longer-term woody crops. The following provides suggestions pertinent to developing such an industry in the prairie region.

Today the establishment costs for afforestation or any large scale tree planting activity are significant. One means to reduce the high cost of crop establishment could be through the participation of an investment equity fund. It would provide equity financing to the farmer to reduce front-end costs. The program would offer one or two equity participation levels sufficient to encourage production up to significant annual acreages. Targeting an end point of 4 M acres (equivalent to 1.6 million ha) in fast growing species, for example in Saskatchewan, will require plantings of 200,000 acres (equivalent to 80,939 ha) per year, which will require several years of establishment. With the life-time costs of approximately $800 per acre (equivalent to $1,977 per ha) to produce a crop, and an equity support mix of 25% and 50%, total support would be in the range of $12 M to $15 M annually after the establishment period.

The fund's position in the crop also will secure future fibre sales. While the farmer will seek to maximize revenue from the sale of fibre, the fund manager would be able to coordinate large scale investment in primary facilities. This would provide a ready market for the lesser-valued portions of the tree (e.g., limbs for ethanol) that may not be readily developed at the farm level.

Government contributions could possibly seed the fund. Farmer participation (overall estimated to be 60% of total fund) would be required. Additional funding would be secured through private sector contributions. Participation in the fund also would provide a ready means to identify, trade and receive payment for carbon credits through a potential trading program. Delivery harmonization would be expected to reduce overhead and increase overall revenues.

Consultations and policy development at the program design level needs to address a number of the key program issues, including defining the enhanced quality and end use aims and attributes of the proposed resource, crop insurance, design of a carbon trading warehouse, and equity fund program design. The risks associated with any new crop development include such things as applicability of farm support programs and crop insurance, to name a few. The combined effect of opening these programs to trees as a crop will build confidence in the farm community and help manage the life cycle risk of the crop (in the same way as any other crop).

The creation of an advisory panel on agroforestry and afforestation would act in a consultation capacity and assist in delivery of all strategic actions. The Panel would be an interdisciplinary group consisting of representatives from agencies active in agroforestry in the province and would oversee the development and administration of programs and policies. This group of experts would lend guidance to program development and to identifying technology and information gaps.

Currently, the responsibility for agroforestry within the Government of Saskatchewan officially resides with Saskatchewan Industry and Resources. However, two other departments have been heavily involved in agroforestry: Agriculture and Food is responsible for the farm based client network and Environment has delivered a program to plant trees on agricultural land. A centralized model of governance for agroforestry would be necessary to ensure effective communication and a coordinated effort for the vision of developing an agroforestry program in the three prairie provinces.

Through a more centralized governance model, developing a policy for access to technology for growers would be a top priority. Timely and wide-spread access would reduce costs and enhance adoption of agroforestry.

Mechanisms to generate more research funding for agroforestry would focus on recognising trees as an important and viable crop for farmers. Some immediate measures could include increasing funds directed to proposal based agroforestry research and development.

8. OTHER POLICIES/PROGRAMS TO SUPPORT AGROFORESTRY

In addition to the review of policy needs, several other types of programs may also encourage agroforestry development in the region. These are described in this section.

8.1. AL-PAC SYSTEM OF PRODUCTION (LEASE LAND)

In 2002, Alberta-Pacific Forest Industries Inc. (Al-Pac) embarked on a large scale poplar farming program in the Athabasca region of north-east Alberta. Their goal is to plant 1,200 ha per year of hybrid poplar over a 20 year period with a long-term goal of eventually having 25,000 ha of poplar farms within a 200 km radius of their pulp mill. These plantations would be expected to provide 12% of the fibre requirements for the mill by 2023. The company has entered into 20 year lease arrangements with farmers and the lease arrangements are based on the distance from the mill: $30 ac^{-1} yr^{-1} (or $74 ha^{-1} yr^{-1}) if located < 100 km radius and $25 ac^{-1} yr^{-1} (or $62 ha^{-1} yr^{-1}) if located between 100 and 200 km radius from the mill. These lease arrangements also include an inflationary clause. Farmers can also earn money by undertaking the site maintenance (cultivating, discing, spraying) for the first four to five years. This company-farmer model currently working in Alberta could certainly be applicable to other parts of the prairies (Al-Pac, n.d.). In addition, the role of trading carbon could certainly aid as a driver for promoting the establishment of plantations across the prairies provided that markets can be developed in conjunction with possible product development. This endeavour by Al-Pac, however, certainly mirrors the early day ideal of planting large scale plantations of trees on the prairies in the 1870s although for different ends.

8.2. MINNESOTA MODEL

Minnesota Agroforestry Cooperative (n.d.) has suggested a plan to reward the tree plantation producers before the harvest of trees. It consists of setting up a Producer Capitalization Fund to finance land for hybrid poplar plantations. These funds are proposed to provide establishment costs, three years of maintenance payments, and an opportunity to producers to advance payments on their anticipated harvest. When the harvest is completed, the Cooperative receives the money which is then returned to the Producer Capitalization Fund to finance future plantations.

8.3. REGULATED CAP AND TRADE PROGRAM

Current opportunities for carbon credit trading using agroforestry projects lie either in the voluntary market or in various GHG regulatory systems. In the voluntary market, the best known system in North America is the CCX. The CCX recognizes a number of agricultural soil and forestry-related project types as being eligible for carbon credits. In the forestry area, the following project types are included:

- Maintaining or increasing forest area: reducing deforestation and degradation
- Maintaining or increasing forest area: afforestation and reforestation
- Forest management to increase stand- and landscape-level carbon density
- Increasing off-site carbon stocks in wood products and enhancing product and fuel substitution

Agroforestry projects may include elements of several of these project types, so that the CCX may provide opportunities for landowners who undertake agroforestry on their lands. The CCX system includes the following rules related to biological sinks projects:

- The project activity involves afforestation on or after January 1, 1990 on unforested or degraded land;
- Eligible afforestation activity should not involve removal of tree biomass, including harvesting or thinning, during the CCX market period;
- Landowners must sign a contract with their aggregators attesting that the land will be maintained as forest for at least 15 years from the date of enrollment in CCX.

Soil and rangeland project types may also be relevant to agroforestry. The CCX recognizes several project types in this area:

- Conservation tillage: Minimum five·year contractual commitment (2006-2010) to continuous no-till, strip till or ridge till on enrolled acres;
- Grass planting: projects initiated on or after January 1, 1999 in CCX eligible counties [in the US] may qualify;
- Non-degraded rangeland managed to increase carbon sequestration through grazing land management that employs sustainable stocking rates, rotational grazing and seasonal use in eligible locations;
- Restoration of previously degraded rangeland through adoption of sustainable stocking rates, rotational grazing and seasonal use grazing practices initiated on or after January 1, 1999.

For agroforestry systems that include intercropping or silvopasture, these project types may provide additional opportunities for landowners.

There are few existing regulatory systems that recognize biological carbon sinks. Canada's domestic Offsets System for Greenhouse Gases (Government of Canada, 2008) is currently under development and will include forestry and agricultural projects when operational. In the forestry sector, projects on afforestation, avoided deforestation and forest management are all recognized. In the agricultural sector, reducing the intensity of tillage operations, adopting crop rotations and grazing management practices that sequester more carbon in the soil, and increasing the use of permanent cover will be eligible. Figure 4 illustrates the steps required for credit creation in the Canadian Offset System.

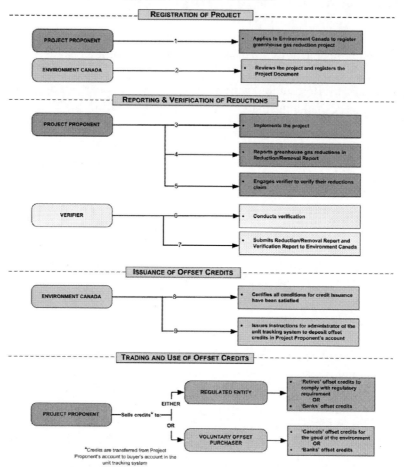

Figure 4. Process for credit creation in the Canadian Offset System for Greenhouse Gases (Government of Canada 2008).

The initial registration of the project lasts for eight years. Due to the relatively slow sequestration rate of sinks projects, re-registration will likely be required. In addition, the removal of greenhouse gases associated with offset credits issued for biological sink projects would have to be maintained by the proponent for a fixed period of time – the "liability period." The length of the liability period has not yet been defined but will likely be between 20 and 30 years for forestry-based projects. There is the potential for loss of carbon from sinks projects during the liability period due to fires, insect attack etc (called a "reversal"). A plan for dealing with reversals will be required for registration. This could include plans for replacing the carbon lost, purchasing insurance against loss, purchasing additional credits from the market to make up those lost from the project, and buffering, in which some credits generated by the project are held in reserve to mitigate losses in the future. In a forest management carbon credit project in Saskatchewan, the historical rate of loss of carbon due to fires and insect attack was calculated and used to determine the size of the buffer required (Lemprière et al., 2002).

The trading of credits will occur in the private sector after credits have been certified by the Offset System and transferred into the proponent's account. An example of a private sector trading platform is the Montreal Climate Exchange (MCeX)[18]. Once credits have been certified through the Offset System, they can be bought and sold through MCeX. Purchasers have the option of retiring the credit by applying it against their regulatory requirements; banking it for later use; or cancelling the credit for the good of the environment.

It is difficult to predict the value of an offset credit in Canada. In a voluntary market such as the CCX, prices have been as high as US$5-6 per ton of carbon but are generally lower. On the other hand, a market associated with a regulatory system such as the European Trading System, in which emitters are required to participate, has seen prices as high as €30-35, although currently prices are €10-15. Price is determined by demand, which in turn is a function of the regulatory system's strength; since the Canadian system is not fully implemented it is currently unknown how strongly it will drive prices upward.

---

[18] Further details can be found in Montreal Climate Exchange (n.d.).

REFERENCES

AAFC-PFRA Shelterbelt Centre. (2007). Hybrid poplar clones for Saskatchewan. AAFC-PFRA Factsheet 01/07.

Al-Pac. (n.d.). Poplar farming. Title. [E-text type]. http://www.alpac.ca/ index.cfm?id=poplarfarming.

Alavalapati, Janaki R.R., & Mercer, D. Evan. (2004). Valuing agroforestry systems: methods and applications. London: Kluwer Academic Publisher. 314 pp.

Atlavinyte, O. (1964). "Distribution of earthworms (Lumbricidae) and larval insects in the eroded soil under cultivated crops". *Pedobiologia, 4*: 225.

Balshi, M. S., A. D. McGuire, P. Duffy, M. Flannigan, J. Walsh and J. Melillo. (2009). Assessing the response of area burned to changing climate in western boreal North America using a Multivariate Adaptive Regression Splines (MARS) approach. Global Change Biology 15: 578-600.

Booth, N.W.H. (2008). Nitrogen fertilization of hybrid poplar plantations in Saskatchewan, Canada. M.Sc. thesis, University of Saskatchewan, Saskatoon, SK.

Brooks, K.N., Gregersen, H.M., & Ffolliott, P.F. (1994). Role of Agroforestry in Sustainable Land-use Systems. In *Agroforestry and Sustainable Systems: Symposium proceedings*. pp. 199-206

Cacho, O. J. (1999). Valuing Agroforestry in the Presence of Land Degradation. No. 99-07. Armidale, NZ: University of New England.

Chranowski, D., & Antonowitsch, R. (1995). *Wildlife Shelterbelt Evaluation in South-western Manitoba*. February. Regina: Prairie Farm Rehabilitation Administration.

Department of the Interior. (1910). Successful Tree Planters. Letters from all over the Prairie Provinces tell of benefits derived from plantations. Government Printing Bureau, Ottawa. 39 pp.

Dickson E., & Fox, J. G. (1989). *The Costs and benefits of erosion control on cropland in south-western Ontario.* Agricultural Economics and Business Bulletin. Guelph: Ontario Agricultural College.

Dore, M., Kulshreshtha, S. & Johnston, M. (2000). Agriculture vs. forestry in northern Saskatchewan. Pp. 148-165. In M. Dore and R. Guevara (eds.). *Sustainable Forest Management and Global Climate Change.* Cheltenham, UK: Edward Elgar.

Dore, M., Kulshreshtha, S., & Johnston, M. (2001). An integrated economic-ecological analysis of land use decisions in forest-agriculture fringe regions of northern Saskatchewan. *Geographical and Environmental Modeling, 5*(3): 159-175.

Dosskey, M., Schultz, D. & Isenhart, T. (2003). Riparian Buffers for Agricultural Land. USDA Forest Service. Agroforestry Notes. Online Reference: Last Accessed May 16, 2003. http://waterhome.brc.tamus.edu/projects/afnote3.htm

Environnent Canada. (2008). *Canada's Greenhouse Gas Emissions: Understanding the Trends, 1990-2006.* Environment Canada, Pollution Data Branch, Ottawa ON. Available on-line at http://www.ec.gc.ca/pdb/ghg/ghg_home_e.cfm, accessed 11 May 2009.

Ellis, J.H., Gill, C.B. & Brodrick, F.W. (1945). Farm forestry and tree culture projects for the non-forested region of Manitoba. A report prepared for the Post-war Reconstruction Committee of the Government of Manitoba by the Advisory Committee on Woodlots and Shelterbelts.

Farness, P. L., & Hesseln, H. (2006). *Agroforestry and water use in Saskatchewan: Examination of the economic and regulatory barriers to accessing water for agroforestry.* March 31 2006. Final Report. Prince Albert: Saskatchewan Forest Centre. 42 pp.

Filius, A. M. (1982). Economic aspects of agroforestry. *Agroforestry Systems, 1*:347-360.

Garrett, H.E., Rietveld, W.J., and Fisher, R.F. (Editors) 2000. North American Agroforestry: An Integrated Science and Practice. Amer. Soc. of Agron. Inc. Madison, WI. 402 pp.

Gordon, A.M. & Newman, S. M. (1997). *Temperate agroforestry systems.* Wallingford, NY: CAB International.

Government of Canada. (2008). *Canada's offset system for greenhouse gases.* Government of Canada, Ottawa ON. Available on-line at http://www. ec.gc.ca/doc/virage-corner/2008-03/526_eng.htm, accessed 12 May 2009.

Government of Saskatchewan. (1997*). The prairie ecozone: Our agricultural heartland.* Saskatchewan's State of the Environment Report. Regina: Saskatchewan Environment and Resource Management.

Government of Saskatchewan. (2005). Speech from the Throne 2005. Available on line at: http://www.gov.sk.ca/news-archive/2005/11/07-1028-attachment.

Griffith, C. (2003). Improvement of Air and Water Quality around Livestock Confinement Areas Through the Use of Shelterbelts. South Dakota Association of Conservation Districts. Online Reference: http://www.sd.nacdnet.org/manure Last Accessed: May 16, 2003.

Howe, J.A.G. (1986). One hundred years of prairie forestry. *Prairie Forum, 11*: 243-251.

Huszar, P.C., & Piper, S.L. (1986). Estimating the Off-site Costs of Wind Erosion in New Mexico. *Journal of Soil and Water Conservation, 41*(Nov-Dec): 414-416.

IPCC (Intergovernmental Panel on Climate Change). (2007). *Climate Change 2007: Synthesis Report, Summary for Policy Makers.* IPCC, Geneva, Switzerland, 104 pp. Available on-line at http://www.ipcc.ch/, accessed 11 May 2009.

Johnston, Mark, & Williamson, Tim. (2005). Climate change implications for stand yields and soil expectation values: A northern Saskatchewan case study. *Forestry Chronicle, 81*(5): 683-690.

Johnston, Mark, Kulshreshtha, S., & Baumgartner, T. (2001). Agroforestry in the Prairie landscape: Opportunities for climate change mitigation through carbon sequestration. *Prairie Forum, 25*(2): 195-213.

Johnston, Mark, S. Kulshreshtha, & T. Baumgartner. (2002). Carbon sequestration on marginal lands: An ecological-economic analysis of afforestation in Saskatchewan. In C. Shaw and M. Apps (eds.). Proceedings of the Conference – *The Role of Boreal Forests and Forestry in the Global Carbon Budget.* Edmonton: Northern Forestry Centre, Canadian Forest Service, Natural Resource Canada.

Josiah, S. (2000). *Discover Profits in Unlikely Places: Agroforestry Opportunities for Added Income.* St. Paul, MN: University of Minnesota Extension.

Joss, B.N., Hall, R.J., Sidders, D.M., & Keddy, T.J. (2008). Fuzzy-logic modeling of land suitability for hybrid poplar across the Prairie Provinces of Canada. *Environmental Monitoring and Assessment, 141*: 79–96.

Kort, J. (1988). Benefits of windbreaks to field and forage crops. *Agriculture, Ecosystems and Environment, 22/23*: 165-190.

Kort, J. & Turnock, R. (1999). Carbon reservoir and biomass in Canadian prairie shelterbelts. Agroforestry Systems 44:175-186.

Kort, J., White, L., & Svendsen, E. (2008). Silvopasture principles and potential in Saskatchewan. *Proceedings of the Soils and Crops Workshop.* Saskatoon, SK: University of Saskatchewan. Feb. 2008.

Kuhn, G., & Nuss, J. (2000). Wastewater Management using Hybrid Poplar. *Agroforestry News.* USDA Forest Service, National Agroforestry Center. Lincoln, Nebraska.

Kulshreshtha, S. N., & Junkins, B. (2001). "Prairie Agriculture, Climate Change and Sustainable Alternatives". Chapter. In H. Haidn (ed.) *The Best of Exploring Sustainable Alternatives: An Introduction to Sustainable Agriculture.* Saskatoon: Canadian Centre for Sustainable Agriculture.

Kulshreshtha, S. & Knopf, E. (2004). *Benefits from Agriculture and Agri-Food Canada's Shelterbelt Program: Economic Valuation of Public and Private Goods.* Research Report. Indian Head: Agriculture and Agri-Food Canada – PFRA Shelterbelt Centre.

Kulshreshtha, S. N., & Kort, J. (2009). External economic benefits and social goods from prairie shelterbelts. *Agroforestry Systems, 75*: 30-47.

Kulshreshtha, S., Knopf, E., Kort, J., & Grimard, J. (2005). The Canadian Shelterbelt Program: Economic Valuation of Benefits. Pp. 247-362. In OECD. *Evaluating Agri-Environmental Policies: Design, Practice and Results.* Paris.

Lemmen, D.S., Warren, F.J., Lacroix, J., & Bush, E. (eds.). (2008). *From Impacts to Adaptation: Canada in a Changing Climate 2007.* Government of Canada, Ottawa, ON, 448 p.

Lemprière, T., Johnston, M., Willcocks, A., Bogdanski, B., Bisson, D., Apps, M., & Bussler, O. (2002). Saskatchewan forest carbon sequestration project. The *Forestry Chronicle, 78*: 843-849.

Letwinetz, B., & Jurgens, A. (2008). Agroforestry demonstration site network. Saskatchewan Forest Centre, Prince Albert. 57 pp.

Lindenbach, Rhonda N. (2000). An Economic Application of Agroforestry in Saskatchewan. Unpublished M.Sc. Thesis. Saskatoon: University of Saskatchewan.

Maisonneuve, C., & Rioux, S. (2001). Importance of riparian habitats for small mammal and herpetofaunal communities in agricultural landscapes of southern Quebec. *Agriculture, Ecosystems and Environment, 83*: 165-175.

Mak, Kevin, Lavoie, Armand, & Grundberg, Byron. (1999). *Evaluating the potential and opportunities for agroforestry in Saskatchewan.* Final report prepared for Saskatchewan Agriculture and Food, Regina, SK. May 12, 1999.

Marchand, P., & Masse, S. (2008). *Issues Related to the Development and Implementation of Afforestation and Agroforestry Technologies for Energy Biomass Production: Results of Focus Group Sessions in Quebec and the Prairie Provinces.* Natural Resources Canada, Canadian Forestry Service. Quebec, QU: Laurentian Forestry Centre.

McKenzie, J. (2003). Saskatchewan's long history of coal mining. Western Development Museum report. 10 pp. Accessed on-line May 19, 2009 at www.wdm.ca/skteacherguide/ResearchDocuments.htm.

McKinnon, Greg A., & Webber, Shelley L. (2005). Climate change impacts and adaptation in Canada: is the forest sector prepared? *Forestry Chronicle, 81*(5): 653-654.

Minnesota Agroforestry Cooperative. (n.d.) Educating Growers and Buyers. Accessed on Worldwideweb on April 13 2009 at: http://rewnewingthe countryside.org/component/option,com_smartpages/task,view/category.70/id,1181/Itemid,43/.

Mitchell, A., (1904). Report of Archibald Mitchell, tree planting inspector. P 20-22 (Appendix No. 5) In E. Stewart Report of the Superintendent of Forestry, Part X, Annual Report 1904, Department of the Interior, Dominion of Canada published 1905, Ottawa.

Montreal Climate Exchange. (n.d.). Accessed at http://www.mcex.ca/index_en.

Peichl, M., Thevathasan, N.V., Gordon, A.M., Huss, J., & Abohassan, R. A. (2006). Carbon sequestration potentials in temperate tree-based intercropping systems, southern Ontario, Canada. *Agroforestry Systems, 66*: 243–257.

Price, G.W., & Gordon, A.M. (1999). Spatial and temporal distribution of earthworms in a temperate intercropping system in southern Ontario, Canada. *Agroforestry Systems, 44*:141-149.

Ramanathan, V., & Feng, Y. (2008). On avoiding dangerous anthropogenic interference with the climate system: formidable challenges ahead. *Proceedings of the National Academy of Sciences*, 105: 14245–14250.

Ross, N.M. (1910). *Tree-planting on the Prairies of Manitoba, Saskatchewan and Alberta.* 4th ed. Department of the Interior, Forestry Branch Bull. No. 1.

Ross, N.M. 1923. *The tree-planting division: Its history and work.* Department of the Interior. 14 pp.

Ross, N.M. (1939). *Tree-planting on the Prairies of Manitoba, Saskatchewan and Alberta.* Dominion of Canada, Department of Agriculture, Publ 623. Farmer's Bull. 64. Ottawa, Canada

Smith, Dean G., & Hoppe, Terrie A. (2002). *Prairie Agricultural Landscape: A Land Resource Review.* Regina, SK: Prairie Farm Rehabilitation Administration, Agriculture and Agri-Food Canada. Electronic version at: http://www.agr.gc.ca/pfra/pub/pallande.pdf.

Sohngen, Brent, & Sedjo, Roger. (2005). Impacts of climate change on forest product markets: Implications for North American producers. *Forestry Chronicle, 81*(5): 669-674.

Thevathasan, N.V. & Gordon, A.M. (2004). Ecology of tree intercropping systems in the North temperate region: Experiences from southern Ontario, Canada. *Agroforestry Systems, 61*: 257–268.

Thomsen Corporation. (1999). *Phase 2 Report, Agroforestry Analysis.* The Thomsen Corporation. Ottawa, Ontario. July 1999.

Tyndall, J., & Colletti, J. (2000). *Air Quality and Shelterbelts: Odour Mitigation and Livestock Production: A Literature Review.* U. S. Department of Agriculture. Project No. 4124-4521-48-3209. Lincoln, Nebraska: National Agroforestry Center.

University of Minnesota Extension. (1999). *Description of Agroforestry Practices in Minnesota.* Access at the Worldwideweb on April 13 2009 at: http://www.extension.umn.edu/diostribution/naturalresoruces/components.

Volney, W., Jan, A., & Hirsch, Kelvin G. (2005). Disturbing forest disturbances. *Forestry Chronicle, 81*(5): 662-668.

Wallace, D. C., Geyer, W., & Dwyer, J. (2000). Waterbreaks: Managed Trees for the Floodplain. *Agroforestry News.* USDA Forest Service, National Agroforestry Center. Lincoln, Nebraska.

Watters, J. (2002). Tree planting in rural Saskatchewan, 1870-1914. M.Sc. thesis, University of Saskatchewan, Saskatoon, SK.

White, L. (2005). *Silvopasture literature review.* Saskatchewan Forest Centre. Prince Albert, SK. 38 pp.

INDEX